DRAFTING

BOY SCOUTS OF AMERICA
MERIT BADGE SERIES

How to use this pamphlet.

The secret to successfully earning a merit badge is for you to use both the pamphlet and the suggestions of your counselor.

Your counselor can be as important to you as a coach is to an athlete. Use all of the resources your counselor can make available to you. This may be the best chance you will have to learn about this particular subject. Make it count.

If you or your counselor feels that any information in this pamphlet is incorrect, please let us know. Please state your source of information.

Merit badge pamphlets are reprinted annually and requirements updated regularly. Your suggestions for improvement are welcome.

Send comments along with a brief statement about yourself to:
Boy Scout Division • Boy Scouts of America • 1325 Walnut Hill Lane, Irving, Tex. 75062-1296.

Who pays for this pamphlet?

This merit badge pamphlet is one of a series of more than 100 covering all kinds of hobby and career subjects. It is made available for you to buy as a service of the national and local councils, Boy Scouts of America. The costs of the development, writing, and editing of the merit badge pamphlets are paid for by the Boy Scouts of America in order to bring you the best book at a reasonable price.

DRAFTING

1982 Printing of the 1965 Revision

BOY SCOUTS OF AMERICA
IRVING, TEXAS

Requirements

1. Prepare the following pencil drawings on material for reproduction:
   a. Make a rough sketch of a room. From it make a finished, scale floor plan. The drawing is to be properly titled. Show by conventional symbols all openings, equipment, and safety devices. List drawing instruments used.
   b. Make a scale drawing of some piece of craftwork. It must be so clear it can be used by someone else to make the article. The drawing is to have a bill of material. There is to be an estimate of cost.
   c. Reduce or enlarge the drawing you made for requirement 1b. Show the scale used.
   d. Show drawings you made of orthographic projections. Show an isometric drawing of something other than those in requirements 1a and 1b.
   e. Make one of the following: (1) A simple bar chart or graph showing your progress in scholarship, physical fitness, or financial activity. (2) A simple schematic drawing of a radio or electronic circuit. (3) A perspective drawing of your own house. Use proper proportion. Show shrubs and trees.
2. Using single-stroke vertical or slant Gothic lettering, describe in ink for reproduction the subject of your drawing in requirement 1b in not less than 25 words. Lettering devices may be used.
3. Describe five ways that drawings can be reproduced. Make copies of one of your pencil drawings and ink lettering using one of these ways.

Contents

Irving, Texas
Library of Congress Catalog Card Number: 19-600
ISBN 0-8395-3273-3
No. 3273 Printed in U.S.A. 3.5M1182

The Language of Industry

As you learn about drafting, you will be getting a lesson in the language of industry. It consists of the lines, symbols, and forms that engineers and craftsmen everywhere understand and can use in making structures or machines from the ideas of other men.

You have probably had the experience of trying to explain one of your ideas to somebody and having him fail to understand what you were describing.

If you picked up the nearest paper and pencil and made a drawing of your ideas, chances are he was able to understand what you were talking about. That is, if you made a clear, simple drawing.

You Are an Interpreter. Just as a language interpreter listens to a person from one country and then gives the message to someone from another country in his own language, a draftsman does the same. He gets the ideas of an inventor, designer, architect, or engineer and translates them as clearly and simply as possible into drawings that are the language of builders and manufacturers.

Simplicity Is Important. It is important to keep your drawing as simple as possible. The simplicity of the lines and symbols allows the sizes of parts of a machine or building to be shown accurately. A simple drawing shows details clearly by bringing out just the surfaces and edges that are essential.

Size, shape, and the position of one part in relation to another are three problems you tackle in every drawing you make. To put in more lines or symbols than are necessary to show these three things would be confusing.

Accuracy Is a Must. An inaccuracy of as little as .0001 of an inch in a dimension on a drawing of an auto crankshaft can result in the scrapping of thousands of these parts. It also stops the assembly line until a corrected drawing is made and new shafts are manufactured.

The answer to questions of accuracy is always to check and double-check. Never allow a doubt to exist in your mind about the dimensions of your drawing.

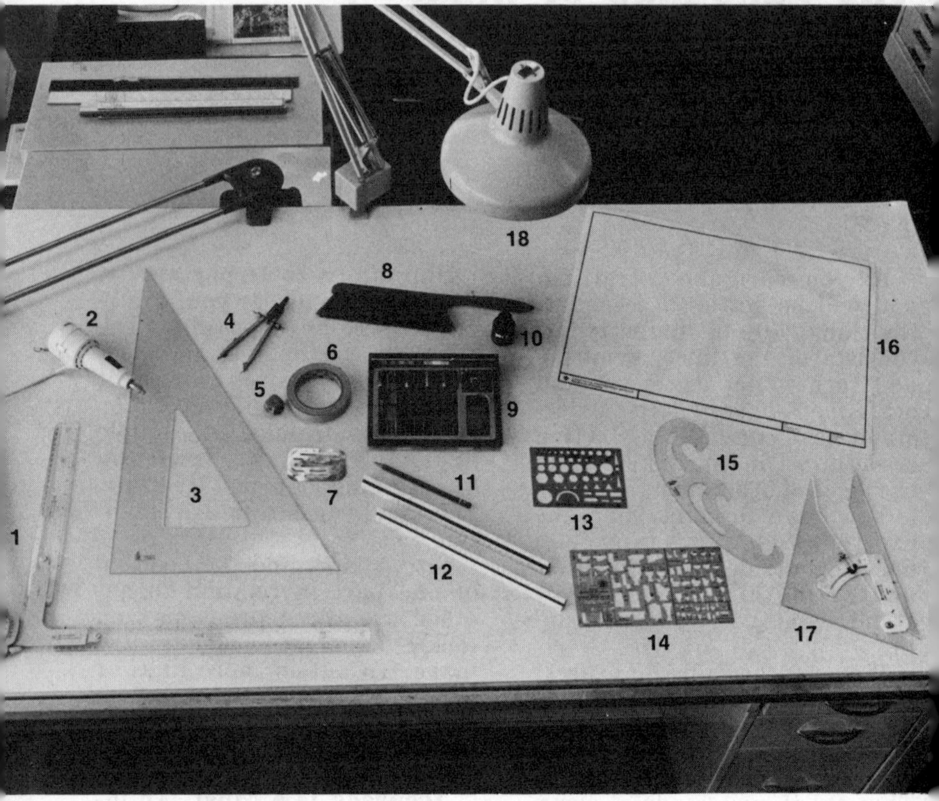

DRAFTING TOOLS

1. Drafting machine
2. Electric eraser
3. 30°/60° triangle
4. Compass
5. Eraser
6. Drafting tape
7. Erasing shield
8. Drafting board brush
9. Technical pens
10. Drafting ink
11. Drafting pencil
12. Architectural and engineering scale
13. Template for general symbols
14. Template for general symbols
15. French curve
16. Drafting vellum paper
17. Adjustble 45° triangle
18. Drafting lamp

Sketching Techniques

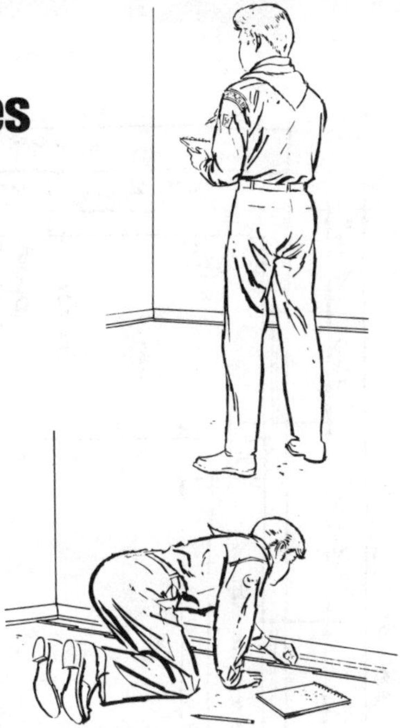

Figures 2 and 3 show a typical rough sketch and finished scale floor plan. The sketch shown in figure 2 was done freehand. The outlines showing walls, doors, windows, etc., were drawn from direct observation, see figure 1a. Then, the room was measured and the sketch dimensioned, see figure 1b.

The information was then drawn to scale using drafting instruments and neatly lettered, making the finished drawing shown in figure 3.

Remember, all floor plans are drawn as though they are being seen from an airplane after the room has been cut off at a level a few inches above the windowsills.

Though technical sketches are done rapidly – not necessarily roughly, depending on your skill –they are more like draftsmen's drawings than artists' pictures, because they show measurements exactly.

The ability to do freehand drawing or sketching is still a valuable asset to a draftsman, particularly in the architectural field and in the field of design. It also comes in handy for studying the best positions of objects before making a finished drawing of them.

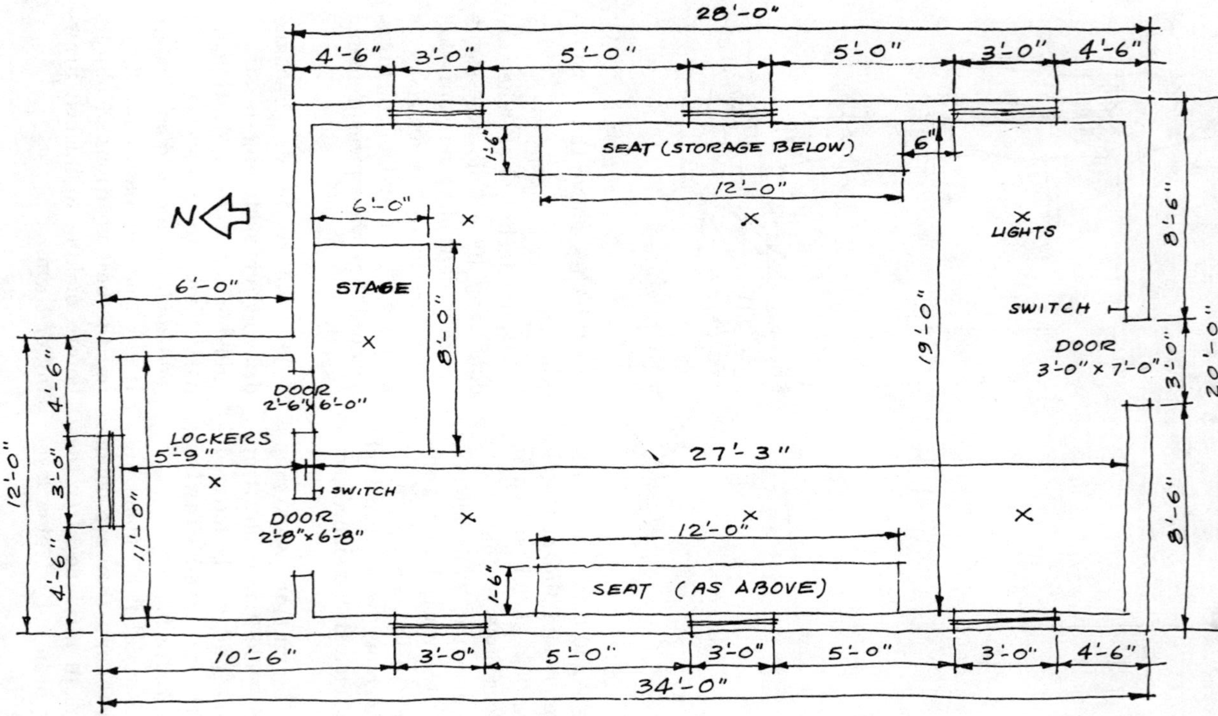

TROOP MEETING ROOM

Figure 2

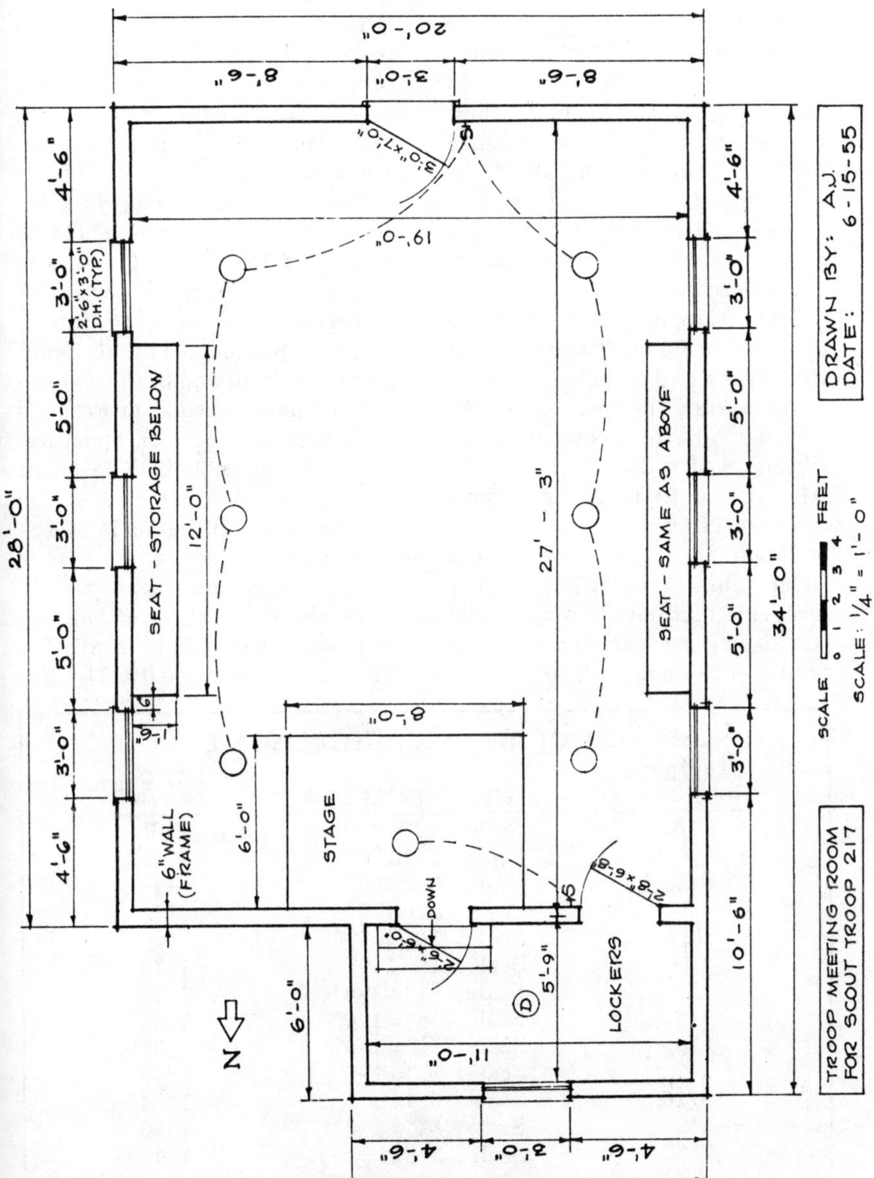
TROOP MEETING ROOM
FOR SCOUT TROOP 217
DRAWN BY: A.J.
DATE: 6-15-55
SCALE 0 1 2 3 4 FEET
SCALE: 1/4" = 1'-0"
SEAT - STORAGE BELOW
SEAT - SAME AS ABOVE
STAGE
LOCKERS
DOWN
6" WALL (FRAME)
2'-6"x3'-0" D.H. (TYP.)
3'-0"x7'-0"
2'-8"x6'-8"
20'-0"
8'-6"
3'-0"
28'-0"
34'-0"
27' - 3"
19'-0"
12'-0"
8'-0"
6'-0"
5'-9"
11'-0"
12'-0"
4'-6"
5'-0"
10'-6"
N

Figure 3

Dimensioning

From the time he first starts sketching the object of his drawing, the draftsman is concerned with dimensions. He must show the actual size of the finished product. Dimensioning should be so complete that the person reading the drawing doesn't have to use mathematics to understand sizes.

For example, in the finished drawing in figure 3 you see that the overall dimension of the south wall of the room is 20′–0″. You don't have to compute it by adding 8′–6″, 3′–0″, and 8′–6″. Also notice that all dimensions are read either from the bottom or from the right edge of the sheet. This is according to the first rule of dimensioning–make the drawing clear and easy to read and to understand.

The Importance of Dimensions

In making drawings to scale it is important to get the right proportions, but it is not good to rely only on the scale for the dimensions of the object that you are drawing.

For instance, if you just indicate this drawing is made to a scale of ¼″ equals 1′, there is a great possibility of error in translating the drawing into a finished product. The drawing parts are small and hard to measure and paper shrinks or expands with heat and cold or moisture. However, do not include more dimensions than are needed for the use that will be made of the drawing.

To assure accuracy, plainly mark the dimensions of the object and letter them neatly so they can be read easily. The fig-

USE OF THE ARCHITECT'S SCALE

MARKED ON END OF SCALE	LETTER ON TITLE STRIP	SIZE OF DRAWING
16	Scale: 12″=1′-0″ or	Full
	Scale: 1/16″=1′-0″	1/192
(No mark)	Scale: 6″=1′-0″	½
3	Scale: 3″=1′-0″	¼
1½	Scale: 1½″=1′-0″	⅛
1	Scale: 1″=1′-0″	1/12
¾	Scale: ¾″=1′-0″	1/16
½	Scale: ½″=1′-0″	1/24
⅜	Scale: ⅜″=1′-0″	1/32
¼	Scale: ¼″=1′-0″	1/48
3/16	Scale: 3/16″=1′-0″	1/64
⅛	Scale: ⅛″=1′-0″	1/96
3/32	Scale: 3/32″=1′-0″	1/128

Figure 4

ures and fractions should be positioned so that they can be read from the bottom of the sheet. Inches should be shown by the symbol (") and feet by the symbol ('). A dash should be placed between feet and inches as follows: 6'–8½".

The dimensions of angles are indicated by an arc inside the angle. Check the examples of dimensioning in the drawings in figure 5.

Figure 5

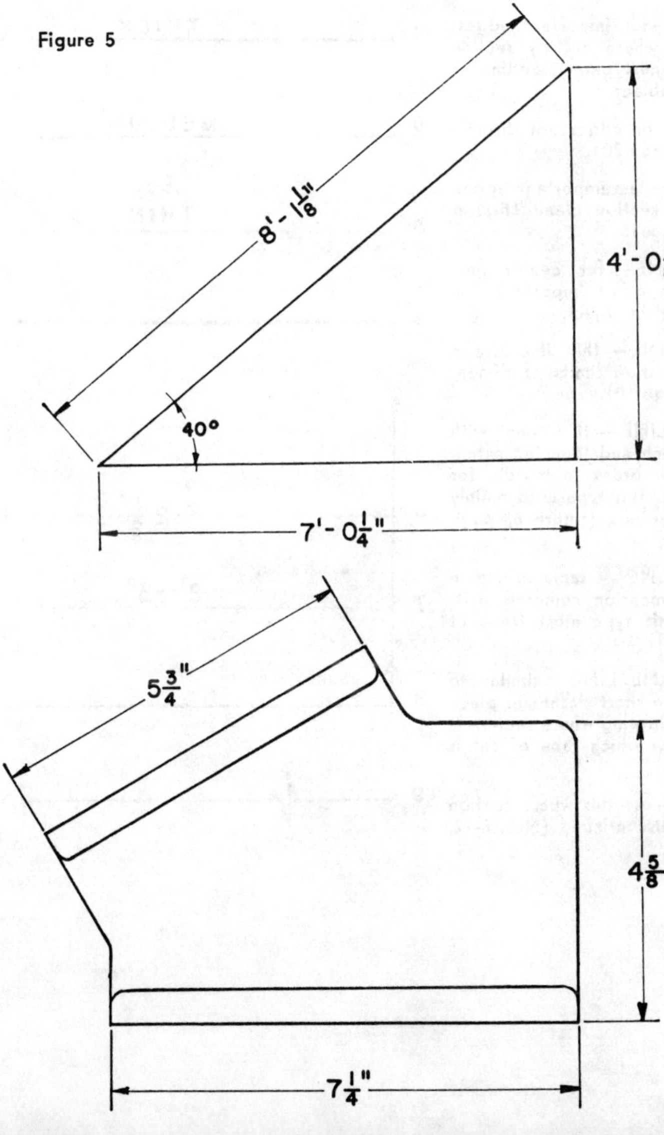

LINES USED FOR DRAFTING
(shown actual size)

1. HEAVY LINE — for important edges: outline of object where cut by section plane (figure 10, next page), outline of elevation view of object.

2. DASH LINE — for edges not directly visible (figure 7, page 20).

3. THIN LINE—for less important edges, or edges beyond section plane (figures 10 and 11, next page).

4. DASH-DOT LINE — for center line, boundary line, or other important line not actually present in object.

5. EXTENSION LINE — thin line drawn free of object to show limits of dimensions (figure 3, page 10).

6. DIMENSION LINE — thin line with arrow points at each end touching extension line and with break in middle for dimension numeral. This type used mainly in mechanical drawings (figure 5, page 13).

7. DIMENSION LINE — same as No. 6 but no break—dimension numerals written *above* line. This type most frequent (figure 2, page 10).

8. CUTTING PLANE LINE — similar to No. 4 but with two short dashes in place of dot. Used for showing where section is taken. Arrows show which face of cut is shown in section.

9. BREAK LINE — use this where section view or plane is discontinued (Nos. 8-12, next page).

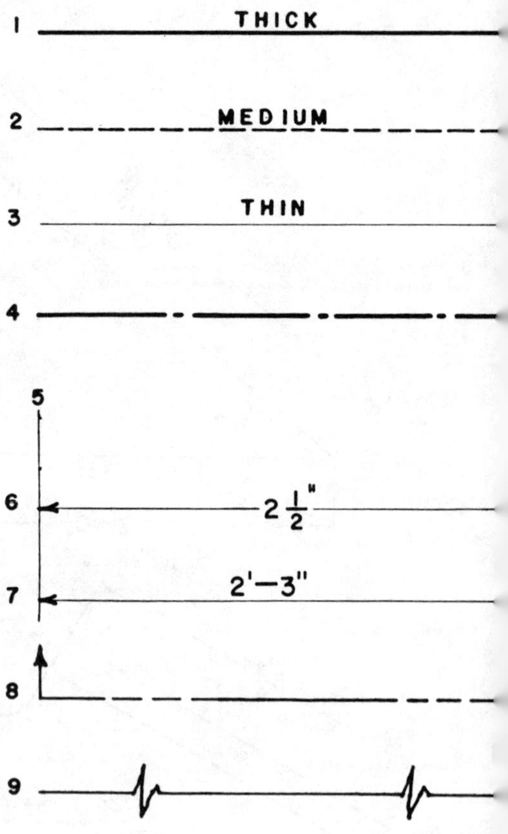

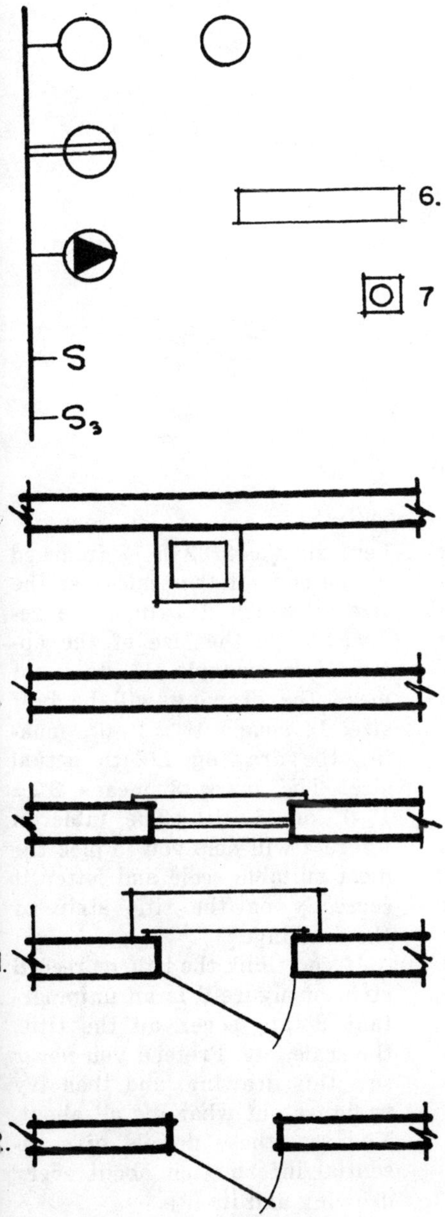

SYMBOLS USED IN ARCHITECTURAL DRAWING

1. Electric Outlet—Wall and Ceiling
2. Electric Duplex Outlet
3. Special Purpose Electric Outlet
4. Switch
5. Switch—Three Way
6. Fluorescent Light
7. Floor Drain
8. Lavatory
9. Wood Frame Wall
10. Double Hung Window
11. Outside Door
12. Interior Door

Scale Drawings

Drawing to scale means fitting your drawing to the size of the paper without changing the proportions of the object, area, or building you are picturing.

Choose a scale that will make the drawing as large as possible without crowding it on the paper. Usually scaling means reducing, but, if the object is very small, it may be drawn full scale or double or more its actual size.

The triangular or flat beveled ruler called the architect's scale is used for all architectural and most mechanical engineering drawings. It makes drawing to a reduced scale easier because the scale does the division of dimensions for you. On the various edges of the triangular instrument 11 scales are given.

Be careful not to take the numbers, such as 3/32 or ¼ (marked at the ends of the scales) as the size to which drawings are reduced from the size of the object. For example, ½ does not mean the drawing will be half size. It means ½″=1′–0″, making the drawing 1/24th actual size. The mark 3 means 3″=1′–0″ or ¼ size. The table in figure 4 will help you to pick the most suitable scale and letter it correctly on the title strip of the drawing.

If you think the title or record strip on figure 6 is an unimportant detail, cover up the title, the scale, etc. Pretend you never saw this drawing, and then try to figure out what it's all about. You see, these details give essential information about every drawing and its use.

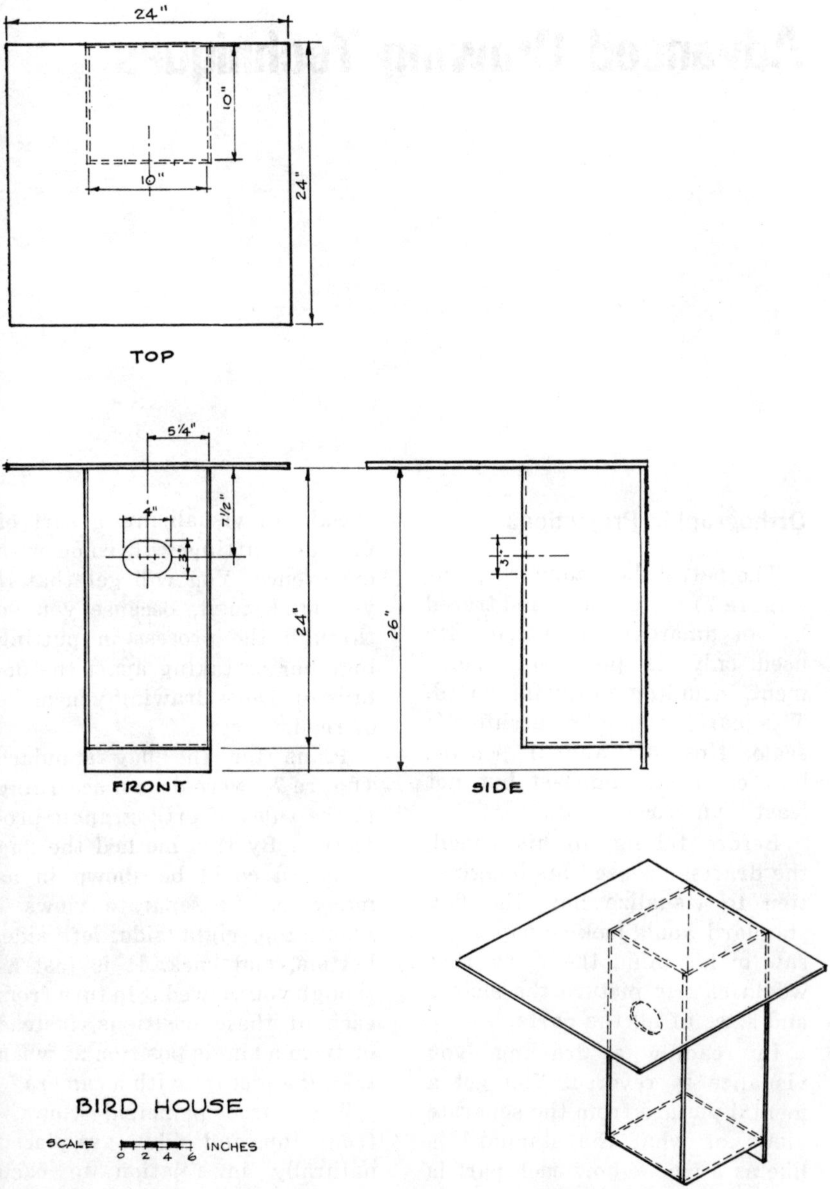
24"
10"
10"
24"
TOP
5¼"
4"
7½"
3"
24"
26"
FRONT
SIDE
3"
BIRD HOUSE
SCALE
0 2 4 6
INCHES

Figure 6

Advanced Drawing Techniques

Orthographic Projections

The patrol flag standard plans (figure 7) were drawn and traced by an amateur draftsman. He used only 11 pieces of equipment, including a drawing board, T square, triangle, architect's scale, Nos. 4H and H pencils, bow compass, and—last but not least – an eraser.

Before taking up his pencil, the draftsman used his imagination to visualize how the flag standard would look on paper. He saw in his mind the views that would clearly picture the shapes and sizes of all the parts.

In reading a drawing, you visualize in reverse. You get a mental picture from the separate views of what the standard is like as a whole, how each part is made, and how all the parts are put together.

Skill in visualizing—a sort of mental drafting—will come with experience. You will get that if you stick to it, because you go through the process in putting together or taking apart the details of every drawing you make or read.

Plans for the flag standard (figure 7) were drawn according to the rules of orthographic projection. By this method the flag standard could be shown in as many as six separate views – front, top, right side, left side, bottom, and back. It is just as though you viewed it in turn from each of these positions, instead of from a single position as when taking a picture with a camera.

The three principal views – front, top, and side – are placed naturally in relation to each other, so that you can tell at a glance which is which without their being labeled. These are necessary to show not only all parts of the flag standard but also their relative positions.

Notice that, although the front and top views contain all the parts, the relative positions of parts are made clearer by a side view. Back and left side views would be only duplicates of the front and right side. The bottom view would not make the drawing clearer.

The three additional views would waste time and material in drawing them plus the time spent in reading them.

Here is the order of work followed in drawing the flag standard. After laying out the sheet with a border line and a record strip, the draftsman decided the three principal views would be enough. These were roughly sketched half size on another sheet of paper. Allowing the customary inch between views, he discovered the scale must be reduced to $\frac{1}{4}'' = 1''$ to make the drawing fit the page size.

Then, the views and a box for the bill of material were blocked in so they would balance on the page. That is, a rectangle with the overall dimensions for each view was drawn lightly. Within these rectangles the views were blocked in, the parts being drawn lightly.

Next, all permanent lines were darkened, and temporary or guidelines were erased. Then dimension lines, dimensions, arrowheads, and lettering were done.

Finally, the drawing was checked for mistakes and omissions.

Requirement 1*b* says that a scale drawing such as this should be "sufficiently clear and detailed to be used by someone else as a working basis for making the article."

Put yourself in that someone else's place – a good way to go about any drafting project. Now, is the drawing complete and clear enough for you to build the standard without asking questions of anyone?

Sure, you'll have lots of questions; but doesn't the drawing alone answer them quickly? If not, it's a poor drawing; or you haven't learned how to read drawings properly.

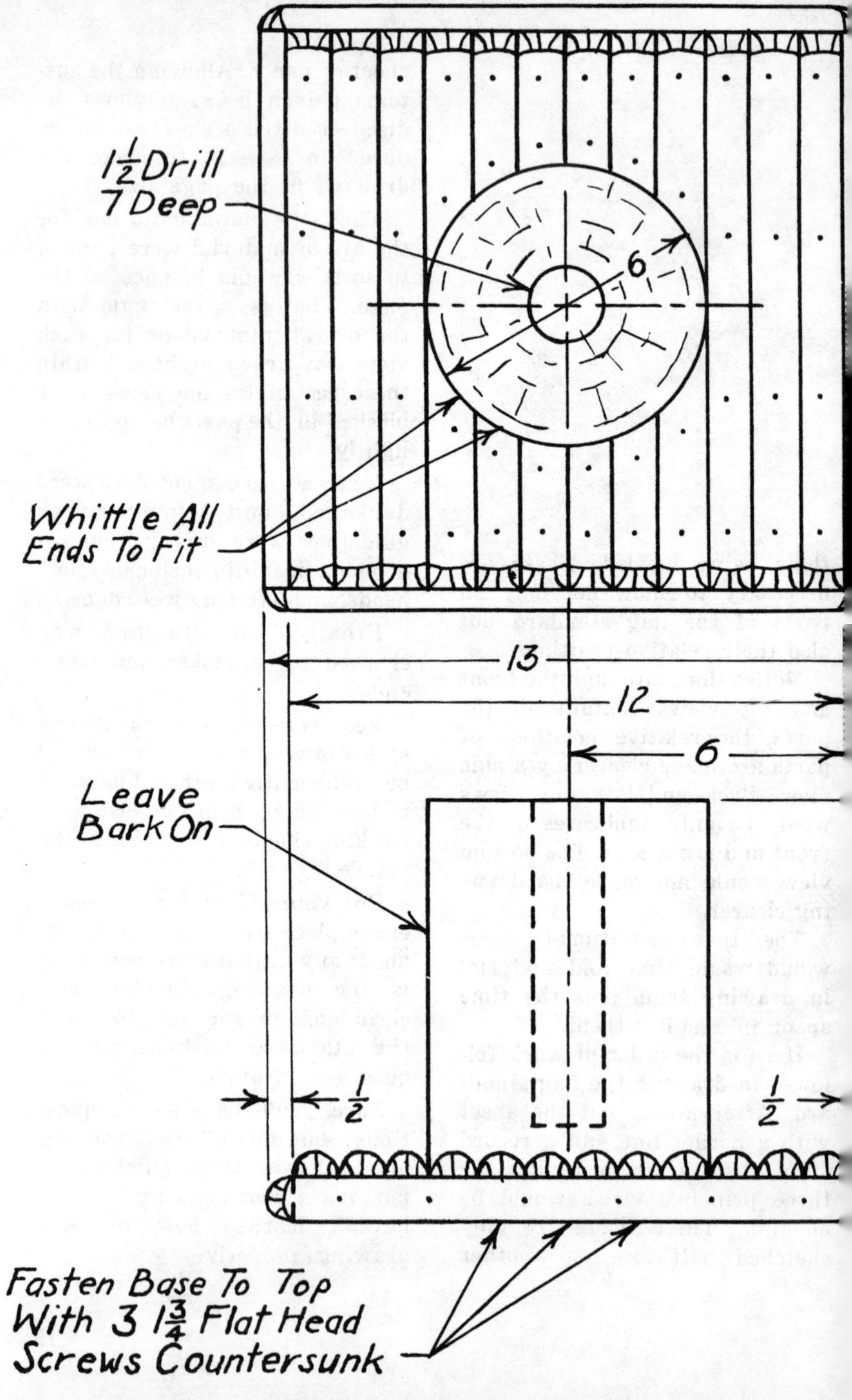

1½ Drill
7 Deep
6
Whittle All
Ends To Fit
13
12
6
Leave
Bark On
½
½
Fasten Base To Top
With 3 1¾ Flat Head
Screws Countersunk

RUSTIC FLAG STANDARD

DRAWN BY JH DATE:

SCALE: $\frac{1}{4}$" = 1"

ALL DIMENSIONS IN INCHES

BILL OF MATERIAL

PART	MATERIAL	SIZE	NO.	COST
Top	Section Of Log	1D×8	1	—
Base	Soft Wood	1×12×12	1	.65
Base	Split Saplings	1D.×12	14	—
Covering		1D.×13	2	—
	Brads	$\frac{3}{4}$	6 Doz.	.60
	Flat Head Screws	$1\frac{3}{4}$	3	.15
	Varnish Or Shellac	4 Oz. Can	1	.50
				1.90

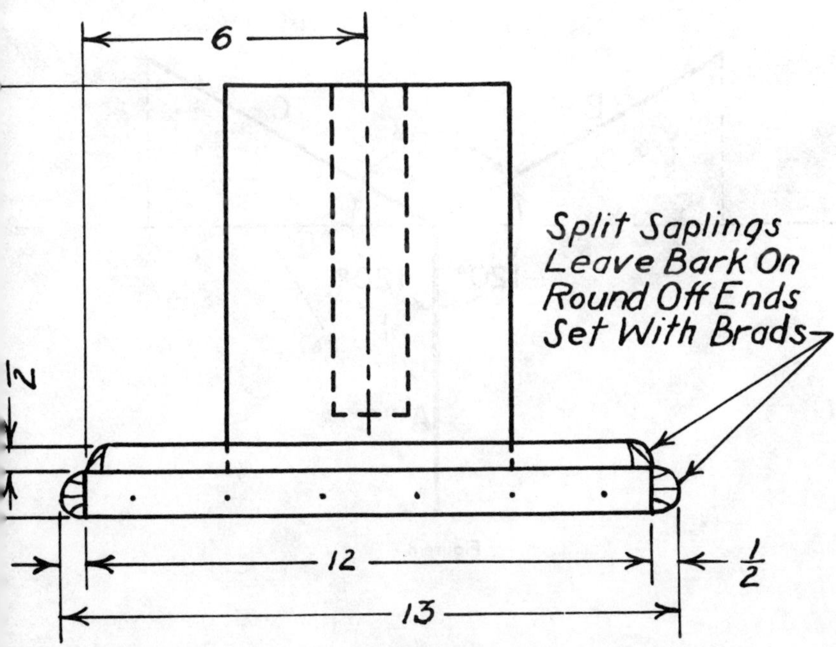

Figure 7

Isometric Drawings

The purpose of isometric drawings is to combine several views of an object in one three-dimensional picture. This picture then can be reduced or enlarged to the size required by means of simple scale measurements.

This kind of drawing is built on a skeleton of three lines called isometric axes (figure 8) forming three equal angles of 120 degrees each. (*From this equality comes the term isometric, meaning equal measure.*)

If you were to draw an object in three dimensions as it looks to you, it would have perspective. The parts farther away from your eye would look smaller (figure 9) than those closer to your eye. This drawing would look realistic, but it would be very difficult to make scale enlargements or reductions from it. The

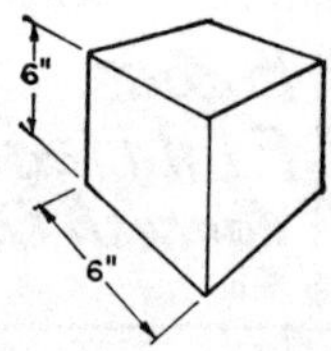

Figure 9

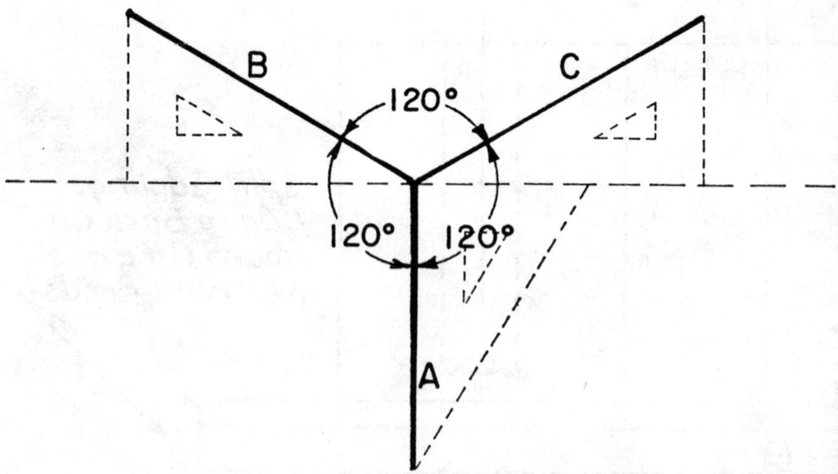

Figure 8

lines farthest from your eye would not give a true picture of their actual length, and you would have to work out proportional measurements that would be hard to interpret to the person reading the drawing.

By building your drawing around the isometric skeleton as you would draw a stick man and then build a drawing of his body, you can make your three-dimensional picture, without perspective troubles. You can then use the axes of the skeleton as "measuring sticks" to make scale enlargements or reductions.

Let's start by drawing an isometric cube (figure 10).

Do you see the isometric skeleton? Notice that all edges of the cube that are supposed to be the same length are actually drawn that way. You can see that all lines that would be vertical in an orthographic drawing are also vertical in isometric drawings, but lines that would be horizontal in an orthographic drawing are tilted upward at 30 degrees from horizontal. (See lines in figure 10.)

If you want to make an enlargement or reduction of this drawing, you can apply your draftsman's scale to lines *a*, *b*, and *c* and convert them to the scale you want in your larger or smaller drawing. You will then have a new skeleton to build on. Try this, and you will understand the isometric principle.

Figure 10

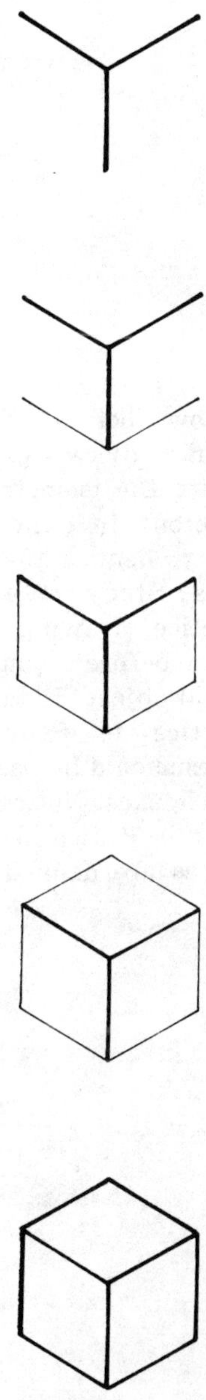

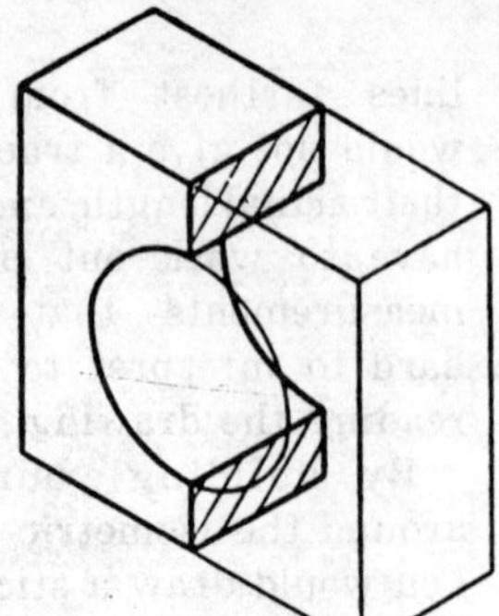

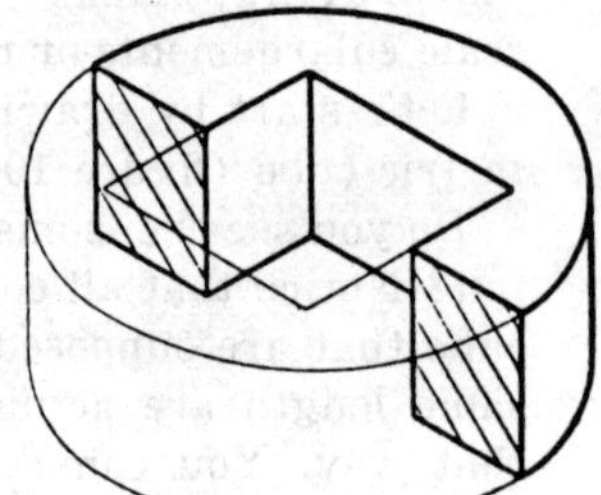

Now look at the other isometric drawings illustrated. Notice the isometric skeleton or skeletons in each of them. See how it defines the three dimensions? Study these drawings and practice drawing them a few times before trying to draw an actual object. It takes additional practice to figure out which edges should be based on the isometric axes. Notice that a framework is drawn first before the shapes are defined.

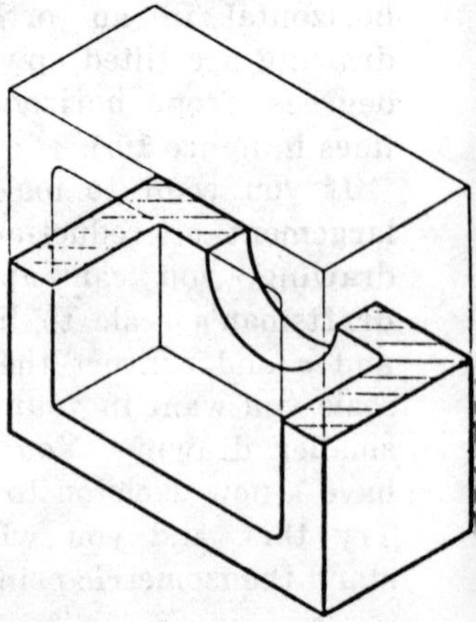

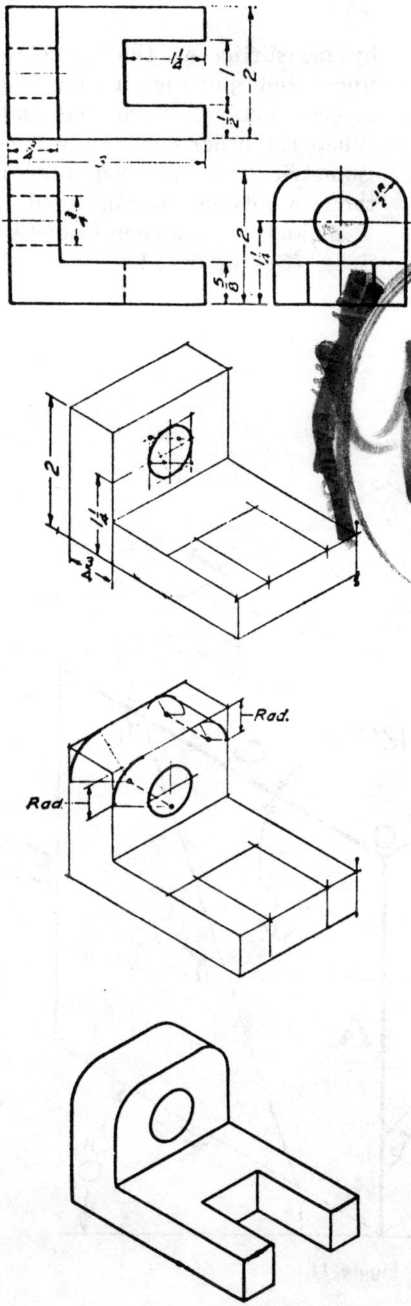
2
2
2
Rad.
Rad.

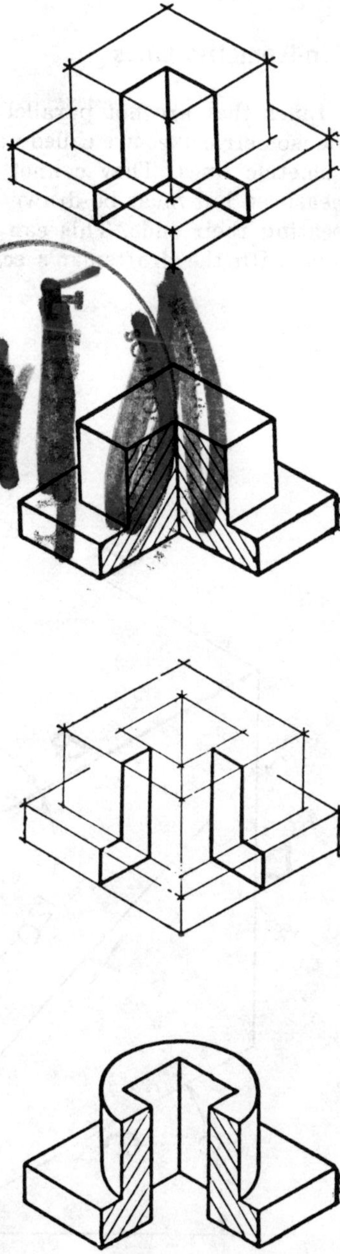

Non-Isometric Lines

Lines that are not parallel to the isometric axes are called non-isometric lines. They cannot be measured but must be drawn by locating their ends. This can be done with the draftsman's scale by measuring on the isometric lines and putting a dot the proper distance from the end. When the other end of the non-isometric line has been located, the two dots can be connected.

X and *y* are non-isometric lines. (See figure 11.)

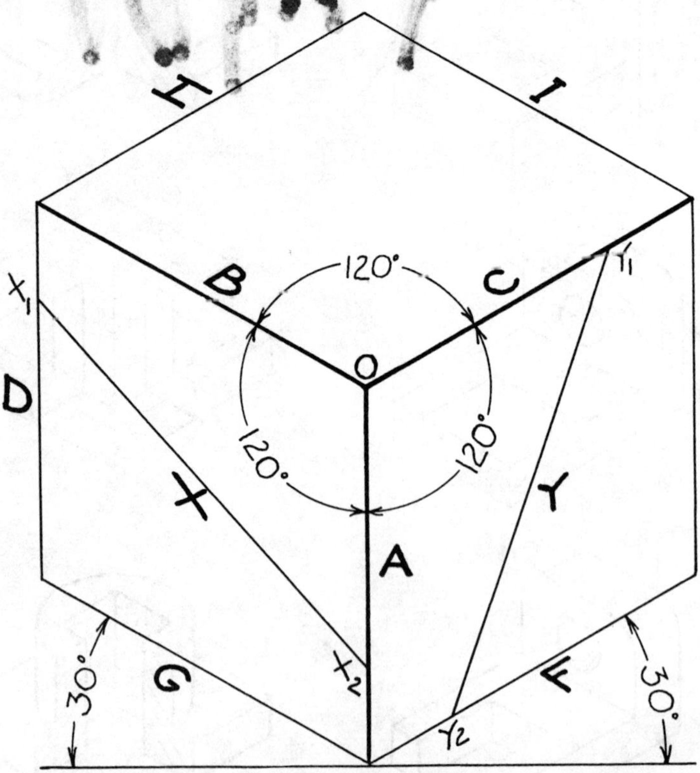

Figure 11

Isometric Circles

To draw isometric circles, the four-center approximation method is most often used. (See figure 12.) Enclose the circle in a true square, then make an isometric drawing of the square. Mark off the points of tangency (where circle and true square intersect) on the isometric square.

At these points draw perpendiculars to the sides of the isometric square, using a 60-degree triangle. The four intersections of the perpendiculars (*a, b, c, d*) will be the centers from which four arcs can be drawn with a compass to complete the isometric circle.

An irregular curve in an isometric drawing is first laid out by the orthographic method on another sheet, and points are marked off at intervals on the curve. The points are located on the isometric projection by measuring the distance of each point from the two axis lines. Finally the points are connected by using a French curve.

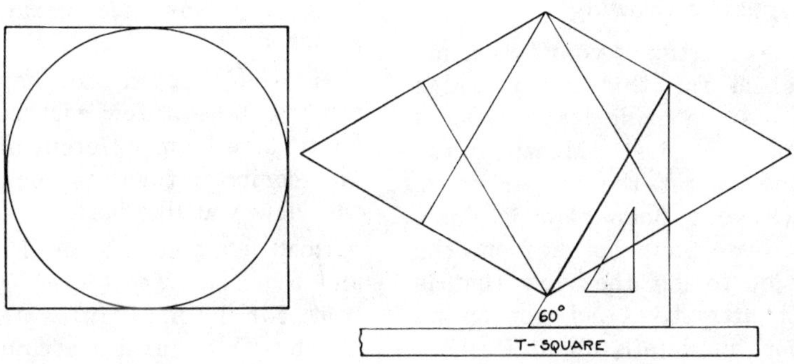

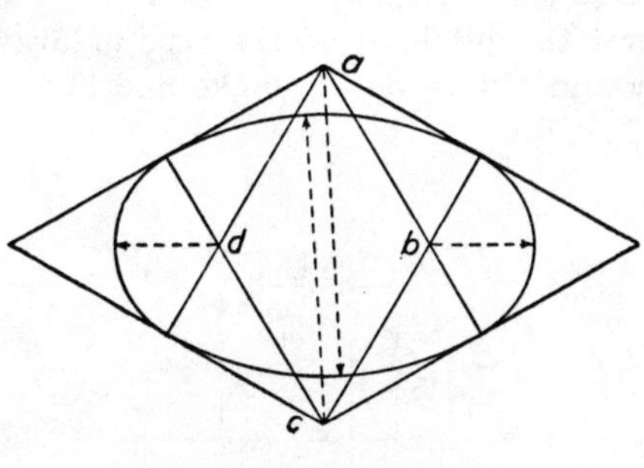

Figure 12

Specialized Drawings

Perspective Drawing

The drawing of your own home required for this merit badge can best be made by drawing a floor plan first. Measure the rooms to get the proper scale. After your floor plan is done, look over your house from the outside to get the view that is most attractive and yet shows important details about its construction. Notice that the drawing in figure 13 shows the pitch of the roof, window and siding details, and how the garage is attached, etc.

If you happen to have a camera, take a few pictures of your home from different angles and compare them to pick out the view you like best.

Read chapter 21 in French and Svensen, *Mechanical Drawing,* for helpful information about architectural drafting and check with your counselor about the type of drawing you wish to make and how to begin.

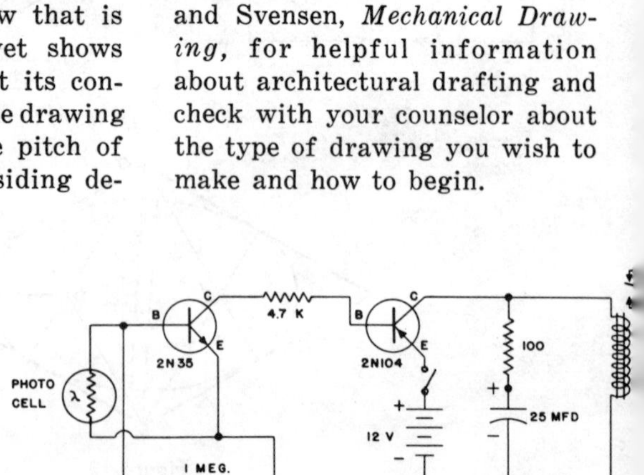

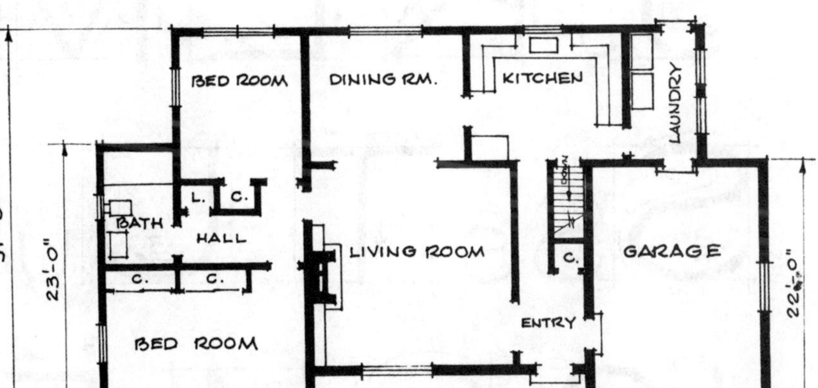

FLOOR PLAN
1/8" = 1'-0"

RESIDENCE FOR: WILLIAM BROWN, SR.
LOCATION: FREEHOLD, NEW JERSEY
DATE DRAWN: JUNE 10, 1965
DRAWN BY: CHARLES BEACH

Lettering

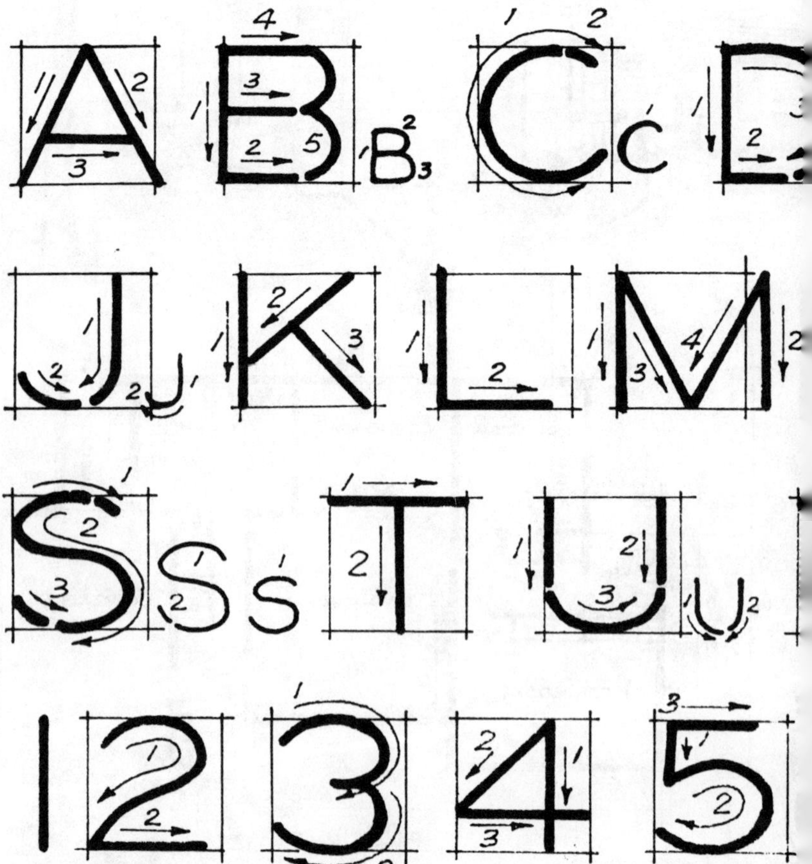

Figure 14

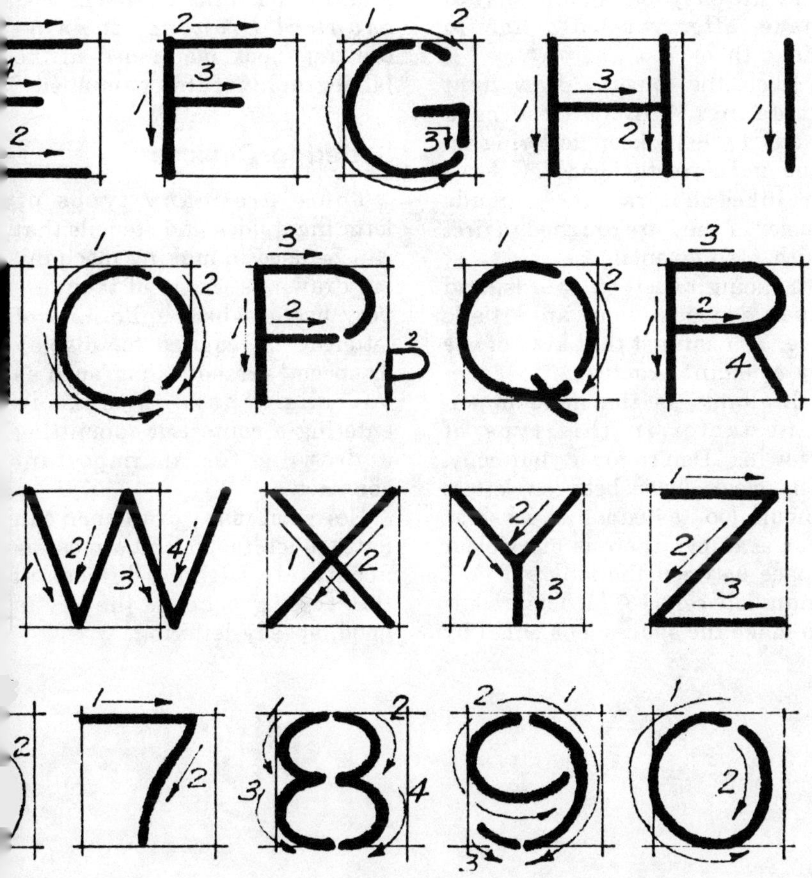

Engineering draftsmen use only the single-stroke Gothic letters, either vertical or inclined (figure 14). Architectural draftsmen use slight variations of the single-stroke roman letters, vertical or inclined.

Either style of lettering may be extended or compressed to give variety or to fit into a certain length of line. The Gothic and roman letters are used because they are easy to read and easy to make, after you learn how to make them.

Even the experts draw light guidelines, which are erased later, to help keep lettering in line and correctly spaced. Colored or inked letters can be made easier of they are roughed in first with a soft pencil.

Spacing of letters, words, and lines depends upon an artistic eye. You can get that kind of eye by constant practice.

Legibility is the more important factor in this type of drawing. Don't crowd your copy. The clear space between letters should look as equal as possible. For example, there is more clear space between the letters T A T than between H A M. The trick is to make the spaces look equal by putting T A T closer together.

When you are through with a block of lettering, you can test the spacing by holding the paper at arm's length and looking for "white spots" between letters. If there are none, you're a good spacer. The ability to letter well and rapidly can be gained only by constant and careful practice.

For further instruction on ink lettering, consult your counselor and French and Svensen, *Mechanical Drawing,* the fine drawing book mentioned in the bibliography of this pamphlet.

Lettering Devices

There are many types of lettering guides and stencils that can be used in making mechanical drawings. Some of them are very useful where uniformity of lettering is required for display or special presentations, such as you might have to make in entering a contest or submitting a drawing for an important contract.

However, many draftsmen can letter much faster and almost as accurately without devices of this type. Practice is the key to good, speedy lettering.

Reproducing Drawings

Blueprints

A blueprint is a copy of a drawing made on special translucent material. The drawing is exposed to a bright light with a sheet of blueprint paper directly behind it. Where the light passes through the translucent material, it turns the sensitized copy paper permanently blue. The lines of the drawing and lettering cut off the light and are reproduced in white on blue.

The sensitive coating of the copy paper is then washed off, dried, and becomes a blueprint.

Vandyke Prints

The vandyke process is similar to blueprinting except that vandyke paper is used, and the resulting print is brown with white lines. It is possible to use vandyke negatives to produce positive prints showing blue or black lines.

Diazo Prints

The diazo process (Ozalid or Copyflex are trade names for it) produces prints exactly opposite to the blueprint process—that is, they are white with blue, brown, or black lines.

Diazo uses ammonia fumes or liquid developer to reduce non-image areas to plain white, leaving dark lines where the drawing protected the copy paper. Diazo prints come out dry, speeding up production, and are easily read and corrected.

Photo Prints

A variety of machines are available to produce photographic copies of drawings. They are useful for making quick prints of details to be incorporated in larger drawings or to be filed for reference.

Disadvantages that prevent photostats from replacing blueprints or diazo prints are their limitations as to size, cost, and use with pencil drawings.

Mimeograph and Hectograph

In mimeograph, the drawing is "cut" into a wax or plastic mat, which is placed on an inked, revolving drum for printing.

The hectograph or Ditto (trade name) process uses a special carbon paper that produces a negative image on the back of a special paper mat as the original drawing is done on the front.

The mat is placed on a revolving drum, negative side out, for reproduction.

Offset Printing

In offset printing, a drawing is photographed and the negative is "burned" onto a chemically treated plate. The plate is then placed on a press where copies of the original are reproduced.

The advantage of using this process is the ability to reproduce enlarged, reduced, or the same size item as the original. Certain areas also can be emphasized by using different colored inks. Printing is an economical method to reproduce large quantities.

Electrostatic Copiers

Most offices have electrostatic copiers for reproducing copies quickly and economically. These machines "xerox" a copy by the action of light on specially treated paper.

However, copies are limited in size, usually to 8½"x11" or 8½"x14".

Careers in Drafting

If you are interested in becoming a draftsman, a year or two of specializing in drafting in high school or technical school may be more to the point than 4 years in college. In addition to drawing, courses in mathematics, physics, woodworking, machine shop, sheet metal work, and any of the other machine and building trades will mean more money in your pocket when you land a job in drafting.

In learning the trade as a drafting office apprentice, a high school graduate had this experience. For the first 6 months he was office boy, getting acquainted with the other employees and equipment. While delivering blueprints to clients, he picked up some pointers on how to deal with people. Next, as filing clerk, he learned the meaning of system when he had to find drawings and tracings quickly.

Finally, he was put at a drawing board but just as a tracer of drawings. Then, as a detailer, he put pencil to paper on some simple parts of machines—routine work. Around the office he kept his eyes and ears open and asked intelligent questions. The draftsmen liked him and gave him valuable tips on his work. Gradually, he got tougher assignments.

At the end of 4 years he was a full-fledged draftsman, able to handle any ordinary job that came to his board. He was working a 40-hour week and earning five times the amount he received when he started.

In another two years a boom in the machine industry increased the demand for experienced draftsmen, so he got a sudden promotion to checker. He checked the drawings of others for drafting errors.

If he learns enough about his

firm's special field, he may become chief draftsman in charge of the drafting room.

Draftsmen specialize just as the companies which employ them do. Here are some of the various kinds of drafting: marine, naval, mapping, aeronautical, architectural, landscape, mechanical, electrical, radio, automotive, plumbing, heating, ventilating, and air-conditioning. The boundaries of these fields overlap. Your specialty will depend on the firm you work for and the training you have had.

If you intend to go further than drating itself, you will want to take college courses including engineering, drawing, or architecture. You may want to go into some other engineering field or art field, using your drafting experience as background.

Books About Drafting

Recommended by the American Library Association's Advisory Committee to Scouting

The three important parts of a mechanical drawring—(1) views, (2) dimensioning, and (3) notes—have been very briefly described in this pamphlet. In order to learn more about the subject, visit your library and borrow one of the books in the following list.

Read the book you selected carefully as it will give you many hints about every phase of drafting, including selection and use of instruments, drafting technique, alphabet of lines, lettering, symbols, and a more complete description of the theory of projection. Work as many of the problems in these books as you can.

Visit the librarian in your school or public library for helpful books and material.

Books

Architectural Graphic Standards, Charles G. Ramsey and Harold R. Sleeper.

Basic Units in Mechanical Drawing. Book I, Randolph P. Hoelscher and Arthur B. Mays.

Beginning Mechanical Drawing Units, William E. Roberts.

General Drafting, Verne C. Fryklund and Frank R. Kepler.

Industrial Arts Drawing and Blueprint Reading, Shriver L. Coover, 1961. A clear and logical presentation of drafting principles and practice for the student.

Mechanical Drawing, Edward Berg, 1962 (revised edition). Defines the subject and develops it completely and graphically.

Mechanical Drawing: A Text with Problem Layouts, Thomas E. French and Carl L. Svensen, 1960 (sixth edition). A comprehensive guide to the theory and techniques of drafting from idea to finished drawing.

The First Book of Mechanical Drawing, Jerome S. Meyer. A simple discussion of the instruments used by draftsmen, the way to use them, and the basic principles of drafting.

Pamphlets

American Standard Drawings and Drafting Room Practices, American Society of Mechanical Engineers, 345 E. 47th St., New York, NY 10017.

Architecture, Art, Drafting merit badge pamphlets, Boy Scouts of America, 1325 Walnut Hill Lane, Irving, TX 75062-1296.

Draftsmen, Occupational Outline No. 7, Science Research Associates, Chicago, Ill.

MERIT BADGE LIBRARY

Though intended as an aid to Boy Scouts and Explorers in meeting merit badge requirements, these pamphlets are of general interest and are made available by many schools and public libraries. The latest revision date of each, which may not necessarily correspond to the copyright date of the pamphlet, is shown below (corrected to January 1, 1982).

Number		©Year
3325	American Business	1975
3388	American Cultures	1980
3398	American Heritage	1976
3395	Animal Science	1975
3381	Archery	1978
3321	Architecture	1966
3320	Art	1968
3303	Astronomy	1971
3324	Athletics	1964
3275	Atomic Energy	1965
3293	Aviation	1968
3313	Basketry	1968
3362	Beekeeping	1975
3282	Bird Study	1967
3378	Bookbinding	1969
3379	Botany	1964
	Bugling (see Music)	
3256	Camping	1966
3308	Canoeing	1981
3367	Chemistry	1973
3253	Citizenship in the Community	1972
3252	Citizenship in the Nation	1972
3254	Citizenship in the World	1972
3390	Coin Collecting	1975
3258	Communications	1976
3338	Computers	1973
3387	Consumer Buying	1975
3257	Cooking	1967
3277	Cycling	1971
3394	Dentistry	1975
3289	Dog Care	1972
3273	Drafting	1965
3206	Electricity	1974
3279	Electronics	1977
3366	Emergency Preparedness	1974
3335	Energy	1978
3376	Engineering	1978
3363	Environmental Science	1972
3346	Farm Mechanics	1958
3348	Farm and Ranch Management	1980
3287	Fingerprinting	1964
3317	Firemanship	1968
3276	First Aid	1981
3307	Fish and Wildlife Management	1972
3295	Fishing	1974
3399	Food System	1978
3302	Forestry	1971
3240	Gardening	1981
3383	Genealogy	1973
3332	General Science	1972
3284	Geology	1953
3397	Golf	1977
3370	Handicapped Awareness	1981
3380	Hiking	1962
3329	Home Repairs	1961
3298	Horsemanship	1974
3358	Indian Lore	1959
3353	Insect Life	1973
3350	Journalism	1976
3355	Landscape Architecture	1969
3389	Law	1975
3310	Leatherwork	1970
3278	Lifesaving	1980
3337	Machinery	1962
3271	Mammals	1972
3339	Masonry	1980
3269	Metals Engineering	1972
3312	Metalwork	1969
3280	Model Design and Building	1964
3294	Motorboating	1962
3336	Music and Bugling	1968
3285	Nature	1973
3306	Oceanography	1965
3385	Orienteering	1974
3372	Painting	1973
3286	Personal Fitness	1971
3270	Personal Management	1972
3281	Pets	1969
3334	Photography	1971
3382	Pioneering	1974
3396	Plant Science	1975
3386	Plumbing	1965
3314	Pottery	1969
3377	Printing	1965
3251	Public Health	1969
3373	Public Speaking	1969
3343	Pulp and Paper	1974
3375	Rabbit Raising	1974
3333	Radio	1965
3292	Railroading	1973
3393	Reading	1974
3342	Reptile Study	1972
3311	Rifle and Shotgun Shooting	1967
3392	Rowing	1981
3347	Safety	1971
3351	Salesmanship	1971
3384	Scholarship	1970
3322	Sculpture	1969
3364	Skiing	1980
3237	Signaling	1974
3250	Skating	1973
3319	Small-Boat Sailing	1965
3291	Soil and Water Conservation	1968
3354	Space Exploration	1976
3255	Sports	1972
3359	Stamp Collecting	1974
3327	Surveying	1960
3299	Swimming	1980
3344	Textile	1972
3328	Theater	1968
3391	Traffic Safety	1975
3371	Truck Transportation	1973
3261	Veterinary Science	1973
3357	Water Skiing	1969
3274	Weather	1963
3265	Wilderness Survival	1974
3315	Wood Carving	1966
3316	Woodwork	1970

Merit badge pamphlets are also available in library-bound editions. For complete information write the Supply Division.